I0428727

HOW I *STRAIGHTENED*

MY SPINE AND LIMBS

by

ETHAN SHERMAN

Copyright © 1990, 1996, 2003, 2011 Ethan Sherman
All Rights Reserved

This book is the story of how I treated myself for a back problem, and my resulting thoughts about the spinal column in relation to deformity, pain and illness. I am not a medical doctor. I am not a healer. I do not treat people for their illnesses. I have no qualifications to do so. Should the reader seek relief for pain or illness, he or she should seek help from a medical doctor.

Please do not assume that even though a straight spine is more desirable than a crooked one, I am encouraging people to try my program. The constant varying degree of pressure which I induced on my neck, and the necessity to snap my neck to relieve that pressure, might make this procedure too difficult for most people to duplicate.

Finally, I am only telling about what I did to myself. I have no idea what the effect of this program would be on other people.

CHAPTER 1

I wrote this book because I believe I made a great discovery. I was able to straighten my spinal column in a unique manner. This was done by working through the senses. There is a connection between the senses and the spine. Nobody ever said this before. If ancient civilizations knew this, it has not come down to us. People who have duplicated my experiments with colors often dismiss it with the attitude that this isn't a new idea because everyone knows that colors affect health. On this point I must differ. Nobody ever wrote about holding their arms in an extended position, staring at a color or viewing sunlight through glass at an angle, and feeling their arms move. This discovery refutes the cliche that there is nothing new under the sun. For that reason alone, it was important to write this book. However, I went beyond this concept, and put the discovery to practical use. I showed how, by pushing against this force, I was able to straighten my spinal column.

The spine is a very difficult subject to discuss. It is perceived as some solid object that is either described as perfect, when no pain occurs, or out, when there is backache present. It is not recognized as a moving, fluid, part of the body, which responds, and is affected, by many stimuli. It is a part of the body that can be fine tuned. Much of what is written about the back is limited to avoiding stress, lifting heavy objects evenly, using the legs instead of the back, losing weight, and exercising to strengthen the lower back muscles.

I try to discuss the spinal column in a different vein. I attempt to show how our back is affected by the work we do, materials we handle, and the sounds to which we are subjected. The two basic experiments in which I turn my arms into a meter can be duplicated in many instances by others.

I want to stress that I am writing a medical autobiography. My program is physically and mentally very difficult. The chronic pressure that I describe, which builds on my neck, and which I induced, bothers me. The fact that I can stop this pressure, discomfort, or even pain, at will, by lying down, does not make me temper my caution. Several times, when I was contemplating writing this book I asked myself if I really wanted others to go through what I am doing to myself. I might have sentenced myself to a lifetime of grabbing my neck and snapping it to relieve the pressure that builds. It is a situation that continues at present. However, the fact that the spine can be straightened is information of the greatest importance. It ranks with splicing genes and splitting atoms, yet cannot lead to nuclear war. I also believe that in a free society all knowledge should be available. In spite of the pressure that builds on my neck, there is a feeling of well being which is hard to describe that I experience as a result of constantly stretching and straightening my spine, limbs, and joints.

I am not a doctor. I am a home improvement contractor. I am not qualified to give others medical advice. Please do not do anything that I suggest without talking to, or getting permission from, a medical doctor. However, I would like those doctors, scientists, and researchers, who are looking for treatments for many illnesses which are spinally related, and for which there is no cure, to investigate my program. They should inquire if Scoliosis could be successfully treated. They should seek to find out if the twisting of the spine and limbs that occurs in Muscular Dystrophy could be reversed. They should ask if the continuous stretching of limbs would enable a Cerebral Palsy victim to discard leg braces. They should seek to find out if the constant stretching of joints and limbs would stop the pain of Arthritis and make joint replacements unnecessary. They should inquire about the effect on Parkinson's Disease, Multiple Sclerosis, Stroke, and paralysis due to spinal chord injuries.

My conception of health is different than current thought. I contend we are a society that is dependent on a medical and drug industry because we destroyed our spinal column through ignorance. We then covered up the injustice through old wives tales. We were told that children are nowadays too busy to stand up straight. People were meant to walk on all fours, so of course they get back pain. We do not exercise enough. We're always, "On the go."

I try to show that this self inflicted punishment to the spine occurs even before birth. It is the most logical explanation for the increase of Crib Death, Birth Defects, and probably even Colic.

The human body is similar to the automobile. It has an intake, and an exhaust system. The heart is like the engine, and the spine is like the frame. If a person turns on his car ignition and absolutely nothing happens, he would not call a transmission specialist, a tire dealer, a wheel alignment company, and then a motor expert. He would realize that the battery was not working, and once it was charged or replaced, everything else would work properly.

It is the same with our bodies. When our spine isn't working properly, we go to one doctor for our foot problems, another for our hypertension and heart ailments, and a different doctor for our constantly upset stomach. We treat the symptoms and ignore the cause.

Just like a car with a bent frame will run, but never run properly, our body will work if it is bent, yet it will never work as it should. The symptoms will be pain and illness in one form or another.

My procedure requires chiropractic adjustments to relieve the pressure on my neck and free the neck to enable the spine to continue untwisting and straightening. Many people have mixed thoughts about chiropractic treatment. My feelings are as follows: The spine, over the course of time becomes misaligned. This can be observed without professional training. Young people get back pain, and older people stoop. When a chiropractor performs a series of adjustments, the individual will still stoop because his spine has been in one position for so many years, that touching it for a few minutes a month won't permanently straighten. In addition, the person will continue the same activities, which, to a large degree, contributed to his condition.

Chiropractors will be the first people who will acknowledge these limits to their field of healing. However, I discovered that by constantly freeing my neck with a series of adjustments, the chiropractor does a tremendous amount of good, in conjunction with my program. That is because by working through my senses

2

I am able to continually straighten my back, joints, and limbs.

Together, I feel we are a terrific combination, not withstanding the difficulties I have with my neck. He is not merely pushing things around, and I am not doing something to an important part of my body without being under professional care. After several months of working through my senses as I describe, it was like pushing an "On" button which activated my spine. My spinal column took off in a positive direction. It was almost like it developed a mind of its own, and knew what it was doing. The mechanics could be compared to a coiled rope untwisting from the top; the top, in this case, being the neck. As the rope untwists it stretches out, but the person untwisting it often has to stop and untangle it so it can continue unwinding. In this manner the chiropractor frees my neck by moving it from side to side, to allow the vertebrea to untwist, and the rope, or the spine, can continue to be released. Incidently, vigorous activity speeds up this process, and resting slows it to a virtual stop. Since the spine does most of its own moving, too many chiropractic adjustments, causing a possible loss of resiliency of the body should not be a problem. However, without the adjustment that I described, and which some chiropractors perform by the famous cracking the neck, as well as two or three other adjustments, the spine winds itself into a poor position. Too much pressure builds on the body, and problems will eventually occur.

In this book I also discuss the change of work we do from a two-handed system to a one-handed method, and the use of artificial materials in today's products. This affects the proper spacing of the vertebrea. I am also concerned about a new type of sound that we hear in the more powerful motors that do our work. Whether they adversely affect everyone, or just those people who can duplicate my experiments, I do not know, but the harmful tuning fork effect of these sounds is something that society will sooner or later have to recognize. While my theory may appear extremely bazaar, it makes sense to me, especially when we consider the unexplainable increase of many illnesses.

During the last generation and a half, a whole series of sounds have come from high-powered engines which have caused a destructive tuning-fork effect on the spinal column. They come from diesel engines, jet engines, FM and CB frequencies, high speed gas and oil burners, hair dryers, copying machines, heat pumps, humidifiers, and air horns, to name just a few.

When men and women relax their bodies and hold their arms in the manner I described, they could sometimes feel the pull or force on their arms. It is often a significant force. Just because the body is rigid in the normal course of events doesn't mean that something isn't pushing against the spine causing an adverse condition. What is happening is a tragedy. It has crept up on us slowly so as not to appear as an epidemic, and there is no national panic, but the results have been disasterous to health.

The most obvious example which has gone undetected is the birth defects and terminated pregnancies by users of video display terminals. These users, and the people who perform the studies documenting the terminated pregnancies, have no idea why this should occur, although they are now beginning to suspect electromagnetic waves. The real reason is the hum of the terminal and its disasterous effect on the fetus.

The worker suffers some discomfort so everyone runs around prescribing eye

3

glasses and better office furniture, but that isn't the answer. This problem is such a graphic example of new sounds adversely affecting the spinal column that it should serve as a modern Rosetta Stone. It should be the clue to make people ask what similar sounds have the same effect.

Our government is preparing to spend over forty million dollars to find out why the black population is becoming more unhealthy. It is no mystery. As Black people expose themselves increasingly to the modern sounds of the work place and home, their spines deteriorate. One of their worst abuses is constantly listening to the FM frequency. This is not to critisize the music. If they listened to it on tape, or on the AM frequency at any volume, there would be no problem. The culprit is the FM frequency.

Bus and truck drivers are physical wrecks. They attribute their problems to being tossed around in their vehicles and having to sit for long periods of time. The real culprit is the diesel engine and sound it emits.

When observing cancer or heart patients, notice how bent and twisted they appear. Look at their mouth and notice that their teeth are often worn down differently on one side of their mouth, which is the sign of a very bad back. While they will all admit to back problems, they will attribute their major problem to smoking, too much meat in their diet, or ancestors with similar problems. Bear in mind that these illnesses were not nearly as prevalent at the turn of the century, when people also smoked, ate meat, butter, cheeses, and heavy cream. In fact, many of these people lived into their nineties. Another reason that makes me believe diet isn't the major culprit is that our pets, who eat extremely well, are suffering from the same diseases as humans. At one time, a mixed breed dog would be very spry at twelve, and live to be seventeen. Now the same type of dog will get arthritis at seven.

It is interesting to me to note that the body can sometimes act as a meter or pointer when in the vicinity of different sights and sounds. When staring at colors at angles, the person's arms move, and, in a short period of time, move with tremendous force. People who have duplicated this experiment do not like to continue it for any length of time. Their usual immediate response to the results of these experiments is that, "It's psychological."

When something moves, it has to do so as a result of a force. Mechanical energy, solar, wind, nuclear, and hydroelectric, are some examples. If the arms move as I described because, "It's psychological," then I have discovered psychological energy. Interested people should ask what else it will move. How can it be harnessed? Will it lessen our dependence of foreign oil?

Interestingly, there are no solid colors in nature. There are only patterns of diversion that cause the eye to continuously wander. The crookedness of the tree, the grain and knots in sawed lumber, the multicolor of birds and flowers, are but a few examples. White, black, and grey, are the only exceptions to this rule, and they don't matter. In a beautiful, well-manicured lawn, weeds appear. Insects and disease attack hybrid plants which are labeled, "Not resistant to disease." What does this mean? Could a force in nature be violently opposed to uniformity, and, as a result, are these insects or plant diseases being methodically sent out to eradicate uniformity? Could this idea be more plausible than the notion that a bunch of hungry bugs set out to eat a weak plant. If the former is true then

4

what is this force, and why does it oppose uniformity? I include this idea even though I am concerned with the spine because when something happens it doesn't only affect one thing. Often, when we search for clues, we must examine an idea from a completely different perspective, and sometime we go off on a tangent.

Please read this book as it was intended. It is an autobiography, an explanation, and a protest. One of the accomplishments of which I am most proud is that this book contains approximately twenty five pages of original thought. This is not a compilation of existing ideas or trendy attitudes. My original manuscript received a great deal of negative reaction.

Whether I was right or wrong was never discussed. Some people even wanted to know what kind of troubled mind would write the things that I did. This attitude saddened me because I thought new ideas would be debated, critisized, laughed at, or even embraced. Ralph Waldo Emerson, in his essay titled Self Reliance wrote, "Trust thy Self," so I decided to take his advice and self-publish this book.

The last chapters are concerned with the sounds we speak and their possible effect on health. Constantly talking in guttural sounds takes a great deal of effort. Perhaps this exertion is our most important exercise. Maybe the way those sounds reverberate against the spine is the fine sandpaper that keeps it very straight. People like the Dutch have long life spans. This is attributed to their great genes. Maybe the credit should go instead to their vibrant speech which is full of guttural sounds. Perhaps the way the Hebrew words were spoken in the Bible up to the story of Babel is the ultimate cryptic message.

This is highly speculative, even by my standards. I included it anyway. The reason was that it represented all I know about the subject. There was a several month period when new ideas seemed as if they were being revealed to me. During that time I was writing as fast as I could to put it all down on paper. Once that period of time ended, there was nothing more I could add.

The explorer Diaz sailed half way around Africa. His voyage lead to the discovery of India by Vasco DeGama. I have taken my program as far as I can. Hopefully, qualified professionals will improve upon it, fill in the missing pieces and put it to practical use.

CHAPTER 2

Several years ago, when I was forty, I developed back problems. I visited a chiropractor who said that I had the most twisted spine he ever saw. After a few treatments he said that I was not responding as I should, and he did not know why. It was then that I began doctoring myself, and through trial and error, developed the following program for untwisting, stretching and straightening my spinal column.

Before going into the details, I want to emphasize some very important points. Everybody who has a health program seems to have been cured, and wants others to follow his or her system. I am not cured. While I have no curvature, after several years of living like a fanatic, I still have a problem with my neck. In addition, I am not looking for people to duplicate my procedure. No one should treat himself without the approval of a medical doctor. Furthermore, I fear that people will find this procedure too difficult, because of the self-induced pressure that builds on the neck, making it difficult to sit in comfort for long periods of time, especially with the back unsupported. I even heard it said that constant neck pain can lead to depression. Also, unlike a diet or exercise program that can be terminated at will, once this process begins, I don't know if it can be stopped. Finally, this program creates a dependency on regular chiropracitc care. Nevertheless there is interest in straightening the spinal column, especially in adults, and without surgery, so I decided to tell my story.

When I began to treat myself, I tried many actions which were futile, and I was very discouraged. I did realize that for some reason my arms were always moving to the right. When I would bicycle, after an hour, tremendous pressure would build on my left arm as it seemed to be pushing very hard against the handle bar. Then one day I held my arms out as if I was riding my bicycle. I noticed that they moved to the right. I remembered that several months before, I had an operation on my left foot. The foot became infected and I was in constant pain. I then started hitting my right foot in the same area where I suffered the corresponding pain in the left foot. I again held my arms out and noticed that they didn't pull to the right anymore.

I then realized that by holding my arms out in front of me in that manner, they acted like a meter or pointer. Next, I asked myself what else would cause my arms to move. I found that the best way to discover the answer was to take a deep breath and relax myself. Then I extended my arms and kept my elbows near my sides. I bent my wrists and loosly pointed my fingers downward.

I tested colors. I stood up, turned my head, and stared at the solid color red placed off to the side. I noticed a pull on my arms. When I stared at red to the right, it pushed my arms away. I made a fist with my right hand and brought my arms back to center. Orange and yellow exerted a pull similar to red but with less force. Now I tried the same experiment with green. I felt the pull which was opposite red, and squeezed the appropriate hand, which, in this case, was the left. I then brought my arms back to center. Later I realized that I could have probably made two fists and clenched them in front of me in the presence of any of this stimuli, but I didn't think of it at the time. At any rate, I put different colors in different positions for ninety minutes or so per day, broken into several intervals, squeezing the necessary fist. I avoided purple. I also stared directly ahead at bright white, such as foam coffee cups. After several months my spine became affected. Pressure would build on my neck. To relieve it I would grab my neck and snap it. I did this about twenty times a day. After a while I found that when sleeping, I could only get comfortable on my side, and would feel my body push against the sleeping surface. Sometimes I would lie on my back and feel myself stretch as in traction. Everyone experiences these things from time to time, but this was done with much more force and regularity.

After I felt changes in my spine, I continued to stare at colors and make fists for several months, but gradually reduced the time, and then I stopped all together. Yet the activity in my neck continues. Here lies my problem. My neck is continuously a preoccupation. There is almost always some degree of pressure, from a lot to almost nonexistant. For all I know it will never stop.

There are several ways I make the difficulty with my neck, which is my only apparent problem, manageable. I lie down. This relieves all pressure, and I am usually fine for a while afterwards. I take a nap as pressure builds late in the day. I go for a walk. I snap my neck. I go for chiropractic adjustments. These visits are followed by tremendous stretching and freeing of the neck and back.

Everybody experiences these feelings in the neck. It is usually so subtle that is isn't readily apparent. When we drive for a few hours we are stiff. That is because the neck is moving but the rest of the body is rigidly encased in a seat and the driver is grasping a steering wheel. Something has to give, and the symptom is stiffness.

It is also possible that the amount of people's neck movement is related to fatigue. People say that they worked so hard that they had to take a nap, or they were so bored because they had nothing to do that they became sleepy. Peoples' necks are always moving, and perhaps the amount of movement is related to the degree of tiredness. Sometimes when a person wakes from a nap he sees lines on his face and shoulders. He will say that he slept wrong. This is merely evidence of a constantly moving neck and spine.

About three years ago I had my only set of x-rays. It showed a four and one half percent twist to the right of the top vertibra, in my neck. X-rays also revealed the top part of my neck veered to the right and the bottom curved to the left. In addition, I have what is described as a military neck. It is straight up and down instead of having the proper curve. On one hand this is quite discouraging. I have put myself through a great deal of difficulty and still have a problem. However, during the four previous years I have felt an untwisting from the base of my spine to my neck. During the time between the taking of the x-rays and this writing, I have incurred tremendous movement in my neck. It moves to the right, the left, it stretches, and then straightens. I lie down when too much pressure builds. Sometimes I fall asleep as different parts of my neck, skull, face, or complete side of my body pushes against the sleeping surface for sometimes as much as thirty minutes or more. When it stops, that area of my body will be red and have lines on it from the pressure my head or shoulders applied against the bed or couch. It feels like a fused or twisted area in my neck is trying to untwist and break free. When one part of my neck appears to have untwisted, pressure builds on the next part. I am encouraged by the fact that while all this activity in my neck is occurring my limbs are continuously stretching and straightening with great force and for long periods of time.

This activity at first only occurred for a few minutes every ten or so days. Now it occurs daily. Often it lasts for long periods of time and is painless. Some examples are while I'm lying down my legs stretch as a foot pushes against an ankle, or maybe I can only get comfortable lying down with my chest pushing against the palms of my hands causing my knuckles to become red due to the pressure. Another time I could be sitting watching television and I could only get comfortable resting my chin in the palm of my right hand, while my right elbow digs into the back of my left hand by the middle knuckle. After several minutes I can no longer be comfortable in that position, and I notice the redness of the back of my left hand where the elbow pushed against it. Wherever parts of my body push against each other or the bed, there is always redness due to the tremendous amount of force and length of time that the activity occurred. There are many variations of the above activity involving virtually all limbs and joints. Hopefully one day my neck will completely unfuse and untwist and all activity in that area will cease. Maybe it won't. Maybe I could stop the activity by reversing the method that started it. For the time being I am content to continue as I have been doing, and see where it leads.

Besides looking at colors, another procedure that had an untwisting effect was staring at sunlight at an angle through a thick pane of glass. This pulled my arms towards the light. Staring at sunlight through plastic pushed my arms away from the light. I then squeezed the appropriate hand, and brought my arms back to center. Incidently, other persons who felt pulls on their arms, have sometimes found them to be opposite mine. Often they were pulled forward or backward. I found that staring directly at mercury vapor lights, which are most of our street lights, pulls my arms to the left, and viewing florescent lighting which is covered by plexiglass pulls my arms to the right. Most people who tried these experiments noticed the same results.

Realizing that the sense of sight affected my spine, I tested other senses. I found that sound has a profound effect on the spinal column. By holding my arms out as in the experiment with color, I was able to notice the following results. Sounds that pulled my arms to the right were most copying machines, some modern cash registers, high powered vacuum cleaners and central vacuum cleaning equipment, welding equipment, FM and CB frequencies, fire house sirens, powerline and mercury vapor light hums, battery chargers, commercial exhaust fans, large dehumidifiers, humidifiers, bathroom hand dryers, tractor, boat, and car diesel engines, video display terminals, stereo television, and automatic teller machines.

Some of the pulls to the left are as follows; Jet engines, train diesels, police and emergency band frequencies, IBM computers, microwave ovens, late model frost-free refrigerators, supermarket freezers, heat pumps, central air conditioning, high speed oil and gas burners, hair dryers, ice-making machines, library

8

detection equipment, bathroom exhaust systems, and impact tools.

These sounds do not offset each other. For example, the force of the diesel truck engine affects the lower part of the spine, while the pull of the citizen band radio is located near the shoulders. By performing the exercise of holding my arms as I described, feeling the pulls of the various products, and making the appropriate fist, my spine also straightened.

Perhaps it sounds ludicrous to suggest that riding on a bus, listening to an FM radio station, or being near a modern heating system with its high speed burner motor is harmful to a person's spine. Yet that is exactly what is occurring. I am sure that this can be proved scientifically.

Animals should be divided into several test groups and fed the same diet. One group should not be exposed to any of the sounds I say are harmful. The next group should be exposed to only one sound. The last group should be exposed to many sounds. The results of this experiment would be dramatic.

Children should be tested the same way. While it might be difficult to accurately control such an experiment, one group of children should be kept away from all the above sounds from conception until eighteen months of age. The next group should be exposed to as many as possible. The general health of both groups should be examined including which group has a larger instance of Colic. This is because back and stomach problems are related.

Doctors are reluctant to deliver babies because of a fear of liability if problems with the birth occur. I don't think doctors are incompetent, nor do I believe that parents are eager to sue. The problem is that these common sounds have caused problems which have made the act of delivering babies more difficult.

I have not written this book to suggest a person spend the day in the office squeezing his left hand to offset the pull on the spine of the copy machine and squeeze the right hand to counteract and undo the harmful effects of central air conditioning. Were that to happen, little work would be done. I wrote my book to point out that what has happened is that in the past four decades we have invented a number of products that heat more efficiently, save us energy, increase work efficiency, and give us instant information. However, the sounds they make produce a tuning fork reaction in the spinal column that harms the back and subsequently undermines health. We have ignored Emerson's law of Compensation.

In the following chapters I describe how the work we do causes the spine to deteriorate. My purpose is not to convince people to adopt a ridiculous life style to keep from getting back pain. The back is made to take a certain amount of abuse, and we should take advantage of that ability and live a normal life. I do try to show though, that there is a right and wrong way to do work as far as the spine is concerned. We inadvertently do so much activity wrongly that we have taken advantage of what the spine can handle, and the result is pain and illness.

It is like the city of Venice. For centuries it was able to handle its pollution by having the tides flush out waste into the ocean. With the large increase in pollution caused by so many tourists, Venice could no longer handle its waste problem and the city suffered.

I try to show how we too are suffering by what we have recently done to ourselves. If I belabor a point it is just to illustrate an idea, not urge people to change their activity. That would be too impractical.

CHAPTER 3

The spine is a difficult subject about which to write. A generation ago nobody discussed it because the back was so mysterious and much information about the spine was unknown. People involved in automobile accidents claimed they suffered back injuries. Whether they did or did not, they were usually awarded about two or three thousand dollars. This was done because it was difficult to prove or disprove anything with certainty.

Nowadays everyone is an expert. People who suffer pain seem to know exactly why it occurred; X-rays show an arthritic condition which will only get worse. A fall caused pain which never went away. Bending caused a pain which still persists. People with back pain believe if it wasn't for the occurrence of a specific event they would live free of back problems.

I would like to offer a different view gleaned from my observations, experiments, and discussions with chiropractors. The spinal column, like other parts of our body, is a superior part of equipment. It can take all kinds of abuse and shock and handle it very well. A person should be able to go through a normal life span pain-free.

To accomplish this there are certain principles of which we should be knowledgeable, and to which we must adhere. When we twist our bodies in a certain manner, especially over an extended period of time, the spine stays shifted. It does not snap back to a correct position. An example is playing the slot machine. After doing this for a long period of time, players report that their body shifts. It doesn't return to a specific position, but the person gets used to the shift, and after a short period of time doesn't notice it.

If a person plays football and runs certain pass patterns, his spine would be different than a person who didn't play football. If a person wears a built up protective shoe after foot surgery his spine would be different than someone who didn't. This is no cause for concern because the human body is made to handle this activity.

To ask a doctor or chiropractor if a person's spine is healthy is a somewhat deceiving question. That is because everyone's spine is different. If a person participates in certain activities his spine might be stretched one way. If he performed in other sports or did different work it would be different. It's almost like asking someone if a person thinks he is good looking. There is no one definition of handsome because everyone is different.

There are, however, methods of everyday activity to which we must adhere if we are to keep our spinal column as perfect as possible. Work must be done with both sides of the body. Until the nineteen fifties work was done that way as a matter of course. A person needed two hands to operate a product. People used these products the way they were intended without giving it any thought, and our spines didn't suffer. When we switched to a one-handed method of work we developed back problems.

The reason for this is as follows; When we stand, our spinal chord is obviously in a vertical position. Our many vertebrae bisect the spine and are horizontal. These vertebrae must be evenly spaced in order to have a healthy spine. When we work with one side of our body, such as using a telephone with the left hand, we pull down a vertebra on the left side. This destroys the proper spacing of the vertebra. When we yank a gear shift lever on our car with our right hand and keep our right foot on the brake, we pull down a vertebra on the right side. When we depress the lever of an automatic can opener we pull down a vertebra on the same side of our body as the hand that we use. It is in this manner that the deterioration of the spinal column and pain occurs. That is why back pain has been built into our society. It is no longer a chance occurrence.

There is a way to restore the vertebrae to their proper parallel spacing. This is accomplished by holding the telephone in the left hand, for example, and stepping on a small object placed under the right foot. A piece of wood or carpet works fine. Using the telephone in this manner eventually pumps up the vertebrae on the left side.

In the previous chapter I showed how my arms act as a meter in the presence of different sounds and colors. There is another way they act as a meter. The reader should try the following experiment. Place your hands together with one palm against the other, and your fingertips touching. Extend your hands forward. Keep your elbows near your sides. Next move your hands back and forth.

Notice that your hands go forty-five degrees to each side of the center of your body. Next hold a telephone to your ear with your left hand. Put down the telephone and immediately try the experiment. Notice your hands will go forty-five degrees to the right, but only about fifteen degrees to the left.

Next, repeat the experiment, holding the telephone to your ear with your left hand while standing on a small object with your right foot. Repeat the experiment. Notice your hands now go forty-five degrees in each direction. If there is a very slight variation it is to allow for the fact that the telephone is made out of plastic instead of a natural material. Again hold the telephone to your left ear and hold an object in your right hand simulating the old fashion way that telephones were used. Again try the experiment. Notice your hands now virtually go forty-five degrees in each direction. When we do work in a manner that enables our hands to go forty-five degrees in each direction, we perform the activity properly, and do not adversely affect the spacing between the vertebrea.

At this point I became somewhat obsessed with the subject. Maybe my theories were correct or incorrect. Yet I was doing things that nobody ever did before, and I found that very exciting. I ask the reader to grant me a certain degree of latitude and exercise a certain amount of patience while reading this book. Everytime I rewrote my manuscripts I tried to edit out this, "Obsession," and each time it crept back in.

I made a list of all activity I performed with only one hand, or activity I did with only one side of my body. I then undid the damage by performing the same activity with an object under the proper foot. This list included mixing food, depressing toaster levers and levers on automatic can openers, brushing teeth, shaving, shaking the shaving cream can, depressing the button on the aerosol can, opening twist off cans, holding a coffee cup, walking with a pen or pencil on my car, etc. Each time I thought of another activity, such as saluting the flag or holding an umbrella, I added it to the list.

I undid the effects of this activity one at a time. I would notice that each time my body would shift after a while. This occurred while I slept. I could tell something was occurring because I would awake in the morning and see lines on my body. Everyone gets these, but these lines were deeper and redder, signifying that great force was being used. Often while still awake, I would feel myself drawn to the sleeping surface with tremendous force. It was sometimes like suction. My hands and fingers sometimes fell asleep. One hand might assume the curled position of a stroke victim. At times I was pressed against my bed with such force that in order to get up I had to gently rock from side to side in order to free myself. Once I considered the effects of the activity undone, I shifted to a two-handed method of doing the work. For example, when telephoning, I hold the cord in the other hand. While brushing my teeth I hold a glass of water in my other hand. When shaving, I hold the razor in my left hand and the shaving mug in my right. Sometimes, at several month intervals I occasionally go back to putting an object under one of my feet. This is because I had been doing certain work one way for so many years that it might take more than several days to completely alter the effect.

What also must occur, and this happens naturally, is that work is never to be done in a flatfooted position. For example, while shaving, using my left hand, not only do I hold the mug in my right hand, but put most of my weight on my right foot. Once an activity is done two handed this happens automatically.

Women are advised by experts not to carry a heavy pocketbook on their shoulders. These experts never go into detail as to just why this activity causes women pain. The reason is that it pulls down the vertibrae on the side that the bag is carried. What women, or even letter carriers must do to correct the situation is put the bag on the right shoulder, for example, and walk with the left foot hitting an elevated object. This can be accomplished by the person straddling a carpet. After a while the person will feel the shift as the vertebrae on the right side assume the correct position.

At one time I remember the question was asked, "Why do Asian people eat with chopsticks?" No one seemed to know the answer. The reason is obvious when we examine the use of chopsticks from a spinal point of view. People stood and prepared their food. This gave them tremendous leverage while cutting. When they sat down to eat with their food already cut, they only had to eat it. It was also significant, as we shall see shortly, that a wooden bowl was used. When persons from the Far East consumed their food they used both sides of their body. That is because they held the chopsticks in one hand and held the bowl in the other.

By contrast the Western method of eating food is almost crude. Cutting the food while sitting at the table, is a very awkward act. Leaving the plate on the table virtually shifts the act of eating from the proper two-handed method of doing work to a one-handed method. This is true even if a knife is held. It is also significant to notice that when using the wok while cooking the body assumes a coiled springlike position. This contrasts to the flatfooted one-handed method of stirring food in Western societies.

Another change for the worse, which occurred in the past forty years is the removal of door knobs and turning levers from doors. Commercial buildings are the worst offenders. Yanking or pushing a door to the side is a very awkward motion. It is offset by twisting a knob. Door knobs and levers should be put back on refrigerators, stoves, kitchen cabinets, etc.

Work is supposed to be done with initial thrust. This is why the single-handle levers on old-fashioned sinks, bathtubs, and stoves, is far superior to round knobs which are easier to grasp but turn evenly and therefore use less initial thrust.

Whenever possible work should be done in a circular manner. A good example would be waxing a car using the circular motion instead of rubbing back and forth. Old-fashioned car door handles with their downward curve do not adversely affect the spine. The modern handles that we awkwardly yank upwardly harm the back. The old-fashioned kitchen sink handles do not harm the spine. The modern single lever sink control that is yanked upward and pushed away is a disgrace.

The way to prove the preceeding allegations as well as the following ones is to perform the experiment which I just described. The work that I said is performed correctly will cause the arms to go forty-five degrees in each direction. The work that I said was performed incorrectly will show a completely different result.

One of the most sacred cows of the back business is that everything should be lifted evenly. This is wrong. Nothing should be done in a symmetrical manner. Lifting evenly, whether it be packages, weights, doing chinups, sit ups, leg lifts, certain yoga exercises, underhanded and overhanded two-handed set shots with a basketball, and two-handed volleyball shots impairs the spine. Each action adversely affects the proper spacing of vertebrea. By performing these activities with an object under the right foot, the vertebrea can be pumped up and the spine can be rebuilt. That is, of course, assuming a person had done these things in the past. Many years ago car hoods, car trunks, and garage doors, were opened by twisting a handle instead of merely lifting evenly. This was a superior method of performing work.

Handling artificial materials harms the spine. While I couldn't do anything about plastic buttons and many small items that are constantly used, I tried to perform some type of compensatory gesture such as placing a small object under my right foot when handling heavy objects with both hands such as aluminum extrusions, vinyl windows, plastic pool products, and swimming pool chlorine. The previous experiment illustrates my contention. Hold a piece of wood or steel. Put it down and move your hands as a meter. Now do the same with a piece of aluminum or plastic. Note the difference.

There is another interesting experiment that I performed that illustrates the difference of the composition of natural and man made materials. I placed a light weight plastic object such as a pen or comb on a table. Then I lifted it. I noticed that I could lift it completely evenly. No initial thrust was needed to elevate the object from the table. I then lifted a stone of approximately equal weight. I noticed initial thrust was needed to lift the stone from the table, and extra restraint was needed to return it to the table, or else it would return more quickly than the plastic object. This means the composition was different, and handling these man-made materials does have an effect on the spine.

The previous experiments of moving the hands back and forth also can demonstrate the effects of smell, taste, and touch, on the spine. Eat a piece of cheese. Move your arms back and forth. Notice your arms will only go about fifteen degrees to the right of center. Drink a shot of whiskey. Notice that the same thing occurs. Smell the flower of a lilac or rose bush. Observe that your arms will go from slightly to the left of center to almost ninety degrees to the right.

The experiment demonstrating the sense of touch is as follows; Shave your face using a blade and dry your face. Move your arms. Notice they go slightly to the left of center, but all the way past forty-five degrees to the right. Now apply aftershave. Notice your arms now go forty-five degrees each side of center.

Try the following experiment regarding sight. It is one you probably won't see on Mister Wizard. Move your hands back and forth, and notice that they again go forty-five degrees to each side of center. Next, repeat the experiment while looking at another person. Notice your hands go forty-five degrees to the left, but not quite forty-five degrees to the right. Now repeat the same experiment while staring at a naked person. Observe that your hands again go exactly forty-five degrees in each direction. Finally repeat the experiment with the person not only clothed, but wearing a hat. Notice that your hands again go forty-five degrees in each direction. The conclusion is that in order to be fully dressed a person must wear a hat.

I believe all these activities that do not allow the arms to go forty-five degrees each side of center damage the spinal column to some extent. This is probably because they destroy the proper spacing between the vertebrea. I am not suggesting a person alter his behavior to take these things into consideration when going about his activities. If someone has to dwell on ordinary work, it would drive him crazy. Imagine squeezing the left hand, or stepping on a stone with the left foot every time a person smelled a flower. I did try to compile a list of activities that adversely affects the spinal column because it should be known. When I listen to one group of commentators say, "Eat right, exercise, get regular checkups, and you are doing all you can do to promote good health," I'm appalled, because they are wrong. They are as incorrect as the other group who says, "Of course you will get pains in your forties. It's only natural." There is much more involved.

CHAPTER 4

There are other abuses of the spine which I want to include; some are obvious, others are not. The spinal column works in such a manner that it can absorb virtually all of these as well as other abuses and the person will feel no back pain nor will he suffer any other ill effects. However, as the back becomes less perfect symptoms will appear. Headaches, foot and knee pain, and arthritis, are a few examples. The abuses of the spine which cause it to be permanently bent one way or twisted another can be reversed. It might not be practical to do so, but the theory should be known.

All sports harm the spinal column. Anybody who plays golf knows that the powerful, viscious, awkward, act of driving a golf ball must in some way be harmful to the back. Yet any sports activity causes us to work with one side of the body. Hitting any kind of ball, throwing a ball, especially downhill from a pitcher's mound, diminishes the spine.

Swinging a bat right-handed over a period of time will leave the back twisted in a certain way. By swinging the bat right-handed while stepping on a small board with the left foot, the spinal column will eventually return to it's correct position. This, like all similar activity would have to be repeated a few times at several month periods. This is because a person might play baseball for many years. Reversing that activity once, for a little while, won't permanently correct things.

The list of such activity is endless. In baseball the basepaths should be run in reverse. If the glove is held in the left hand, the person should step on something with the right foot. Humpty Dumpty can be put back together again.

I find that it is important to live in the real world. My son and daughter like to play tennis and football with me so I play those sports even though they are some of the worst offenders. On the other hand, I've made a list of sports in which I no longer participate and have reversed what I consider to be the ill effects. This includes dribbling and shooting basketballs, hitting golf, billiard, and volleyballs, while stepping on something with my right foot, since I am left-handed. This may sound rediculous. Yet nothing, in my opinion is sillier than chinups, situps, pushups, lifting weights, etc. However, these activities are socially acceptable.

I used to do these exercises. Now I consider them a waste of time. I believe they ultimately do more harm than good because they are activities performed evenly. While we must avoid a sedentary life-style to maintain health, we do not have to exercise to be healthy. These exercises are something we do not have to do in the normal course of events. For that reason we eventually stop exercising, and then we feel badly about losing all the tight muscles we developed. For all I discuss, I only do what I can perform in the normal course of events, and after an extended period of activity, I now only do things I can perform without thinking. For example, when I answer the telephone, I automatically step on the carpet. However, I won't stand outside a shopping mall waiting for someone to yank open the door. When I jog it is because I enjoy the activity, not because I need to do so in order to acquire health.

Another activity that harms the spine is the act of not reversing our paths. Just like we run the basepaths by constantly making ninety degree turns, we repeat the same principles in our everyday patterns of life. We leave the dining room table, clear off our plates by the garbage, go to the sink to deposit the dish, and finally leave the kitchen. We use the toilet, then walk to the sink to wash our hands, then leave the room. Our banks, supermarkets, municipal buildings, etc., have a procedure to handle the traffic whereby we are ushered in one door, follow a prearranged traffic pattern, and exit by a different door. In each instance the spinal column twists a certain way and stays twisted. These paths should be reversed.

The eye is a powerful muscle. If a person uses one eye or looks in one direction, the other side of his body should be put to work. In military drills, when the command, "Right face," is given, extra weight is put on the left side of the body. As the soldier looks to the right he twists his right foot to the right, thus taking the weight off it. Next he puts his left hand on his left hip which puts more emphasis on the left side. This is not much of a practical problem because we usually do this type of activity automatically. However, we should be careful to observe this principle while watching television off to the side, or viewing objects through microscopes and telescopes, especially with one eye closed.

There is a very serious abuse of the spine of which we are not aware. We often breathe incorrectly. Human beings are supposed to breathe by inhaling through our noses and exhaling through our mouths. Breathing in and out through our mouths not only impairs our spine, but does not let us take advantage of the filtration system which is located in our noses. We would never drive our cars without a gas, air, and oil filter. We wouldn't operate our air conditioner without the proper filter. We wouldn't use our furnace without the proper filter, and neither should we breathe without using the proper filtration.

I recently heard a doctor who specialized in breathing disorders discuss the rise in lung diseases. He and the audience naturally assumed the reason was the increase in air pollution. This is not necessarily so. The reason is more likely that as our spines worsened, we lost the ability to breathe properly. Perhaps this misconception will ultimately benefit our society, as it will cause us to try to stop the air polluters.

We should never get only one hand hot or cold. In the past, dishes were washed in a full basin of water. When men shaved, they filled the sink with water. In both instances the person's two hands were immersed in water. Doing work in that manner took more time, but was worth it. When Asiatic people ate, they held a wooden bowl. Wood was used because it wouldn't transfer heat. Masons work very hard and suffer injuries. Yet they do not do themselves any good, when, at the end of the day, they grab a hose and wash one hand, then switch hands and wash the other.

Finally, the spine has a couple of curves in it. One is by the lower back and another is in the neck. Notice when sitting in a theater where the seats are quite old, that they are virtually straight. There is little padding to support the lower back. That is because many years ago it wasn't necessary. Today it is. A few years ago, when someone went to a movie, he could sit up or slouch in his seat. He had a choice. In many new theaters seats are so constructed that the patron is forced to slouch. Bucket seats in automobiles force the driver or passenger to pad his lower back. This is unfortunate because if everything was working properly, the arch on a person's back or even his feet, for that matter, should develop naturally, and not need padding. This padding is a crutch, and does the work the back and foot are supposed to do themselves.

Since embarking on my program, which involved doing almost everything I described, I have noticed several changes in my body. Some are dramatic, and others seem insignificant, but they all seem to be spinally related. When I switched activity from one-handed to two-handed and undid all the one-handed actions I could remember, my back pains disappeared. Then I began working through my senses. After a few months, I felt pressure build on my neck, and began snapping it to relieve that pressure. A few weeks later, I began to feel my body shift and stretch while lying down. At first this would happen about once every two weeks. A knee, the top of my foot, and my face would seem to laminate against the bed while I slept. At first this only lasted a few minutes. After a while my body would shift every few days. Presently I feel my body push against the sleeping surface almost daily. This sometimes occurs for twenty or thirty minutes at a time, with varying degrees of force. It is very dramatic. Picture one person grabbing your ankle and another your knee and pushing the outside of your leg against a bar in order to straighten your leg. Imagine someone grabbing the top of your foot and pulling it downwards with great force for ten or fifteen minutes in order to straighten your ankle. Picture someone putting a board on the top of your outstretched hand, then sitting on that board for five minutes, straightening the knuckles on the back of your hand. These are just a few of the types of activity I experience as my body stretches. It is awesome and painless.

After a few years of this activity, changes occurred. My stomach problems ceased. Hiccupping virtually stopped. Biting my mouth and tongue very rarely occurred. That sensation known as food, "Going down the wrong pipe," very seldom happened. The very few times any of the above took place was when there was a great deal of pressure on my neck, on a few occasions, when I felt a great deal of pressure on my neck, I did have a little trouble swallowing food.

Occasionally, In the past, I would sometimes awake and find that my hand was, "Asleep." This no longer was a problem. When I jogged, my foot would at times twist for no apparent reason. This also stopped happening. Footwear manufacturers are trying to devise a basketball shoe that will stop a player from twisting his ankle even when no contact is made. This will be futile because they do not understand that the condition is spinally related. If they do invent a shoe that works for that purpose, it will probably wreck the athlete's normal walk. Another change that I noted was that as my neck kept untwisting I often felt my chest pop, as if something constricted was being freed. This made me appear more broad shouldered.

About five years ago I developed mild allergies for the first time. Approximately two years ago they ceased. I can't say for certain why they stopped. My guess is that over a period of time my face constantly pushed against my bed while my body stretched. I would awake, look in a mirror, and see my forehead, nose, and cheekbone, red from the pressure. I believe that after a while, my sinus passages stretched to the point where they worked better, and everything could drain properly.

On the other hand, by constantly having my face push against my bed in that manner, one of the lines on my face, the one that goes from the bottom of the nose to the side of my mouth, became somewhat more pronounced.

My eyesight and hearing were somewhat affected. I do not wear glasses. I noticed that when I sat in the diner having coffee and reading the paper, I had to hold the paper farther and farther away. That condition began to change. My eyesight got better, then changed for the worse, and then improved again. I can now read the newspaper at the diner while holding it at a normal distance. I realize that I could read it more easily with eye glasses, but if I do wear them I am hooked for life. I am hoping that as my spine continues to straighten, my sight will keep improving.

My hearing must have been slowly deteriorating without my realizing it. Several years ago the word, "Huh?" began creeping into my vocabulary more and more frequently, when people spoke to me. During the past few years, I stopped using that response. Recently, I had my hearing checked by an ear doctor who said that it was fine.

It is important the way our country views health. Fifteen or twenty years ago almost anybody interested in nutrition as a method of enhancing health was considered a food faddist and scorned. Doctors were the biggest opponents of the idea that extra concerns about nutrition were valid. Now the medical establishment has done a one hundred and eighty degree turnabout. It now believes that most of our physical ailments stem from a poor diet, and for that matter, lack of exercise. What is somewhat disgraceful about that attitude is that these people act as if they originated this idea and completely ignore the early pioneers who said the same things more loudly and elequently. Adele Davis, Nathan Pritiken, and J. l. Rodale, are just a few of the obvious names which come to mind.

Proper nutrition is important. Eating whole grains and lightly cooked fresh vegetables is a wonderful idea. It is also important to avoid aluminum and flouride in the diet. The latter washes the calcium out of the system. Espousing good nutrition was a stroke of genius for the medical profession. It placed a large portion of the responsibility for health on the patient, and relieved doctors of an equal amount of blame. A doctor could say, "Of course he is sick. Look at all the hot dogs he ate during the past thirty years." Since many people feel guilty, the family would hang their heads and say, "Yes, we have sinned."

Diet is not the major abuse of our body that has lead us to a life of pain, and made us dependent on a health plan to survive. It is not the reason we are seeing so much heart disease and cancer. The real culprit is the undetected self-abuse of the spinal column. The way we are forced to do work; The artificial materials we must handle; The sounds to which we are exposed; are our greatest enemy. We see evidence of this in the increase of back pain, and the corresponding rise of illness. Much of this sickness might not seem to be related to back pain, but it is.

It is important not to get sick. This may sound simplistic, but when someone needs medical attention for an illness, he never only has one problem. The person with a rapid heartbeat might take medication for that symptom, but soon he will develop high blood pressure, then maybe gall bladder attacks. Perhaps he will suffer with prostate troubles, or develop a hernia. It is no coincidence that one ailment will follow another. That is because everything is symptomatic.

Hemorrhoids is an example of a condition that never occurs alone. There is a famous baseball player whose suffering with that condition made the newspaper headlines a few years ago. Since that time he has had numerous injuries such as pulled muscles. These injuries were referred to in the media as puzzlements. It is really no mystery. Hemorrhoid problems are the sign of a third rate spine. When you ask a third rate spine to do first rate work, something has to give. That is why muscles tear and hamstrings pull.

Recent studies show that the people who are benefiting most from a low cholesterol diet are not dying from heart disease, but nevertheless are not increasing their life span. This is very sad, but it is also understandable. When the spinal column is deficient, illness occurs. Some people feel a sense of comfort that a relative who died, and who adhered to a strict diet, was shown as a result of an autopsy, to have very clean arteries. However, if illness doesn't appear in one form it will show up in another, even though cause and effect are not apparent. When a person says that he is in perfect health except for the standard back aches, it's almost like saying, "Otherwise Mrs. Lincoln, how was the play?"

As I mentioned previously, I must regularly go for chiropractic treatment because of the mechanics of my program. The vertebrea in my neck are twisted to the right. Pressure on my neck builds because it always feels as if it wants to move to the left and untwist. When this pressure builds to a great degree, I grab my neck and snap it to relieve that pressure. After a while it has to be relieved in another manner. That is the job of the chiropractor. I lie relaxed on my back, and he stands behind my head. He takes my head in his hands and gently moves it back and forth from left to right. The main part of this adjustment is moving my head to the right. It is physically impossible for me to do this. Some chiropractors perform this adjustment by cracking the neck. That is an action that has brought the profession a lot of notoriety. However, the adjustment can also be done the way I described.

For several years I went for adjustments every two or three weeks. Then something terrible happened. My medical plan refused to pay for any more chiropractic visits. They said that I was going for regular maintenance instead of treating a specific problem. I believe I could have successfully appealed had I had more x-rays taken. Actually I would have liked to have known what changes had occurred in my neck in the three years since my first x-rays were taken. However, there is a risk involved with x-rays, so I didn't bother.

What I did was teach my wife how to perform the adjustment. She did a great job, and I was able to space my regular visits eight weeks apart. There are a couple of other very necessary adjustments that the chiropractor performs. These were difficult to explain, so I didn't even try.

I would like to see those professionals who concern themselves with healing investigate my procedure to see if it has merit. They could do a lot worse. Cerebral palsy victims used to endure painful corrective surgery which proved to be useless. Perhaps researchers would find that if the victim could straighten his limbs by himself by working through the senses, the disease would be cured.

Joint replacements are becoming a more common operation. Besides the obvious pain and risk involved in an operation, these joints often have to be periodically tightened by surgery. These procedures, which are in their infancy, rely on glues and plastics which are still being perfected. Moreover, a joint just doesn't wear out. It has to go bad for a reason. That cause is the severe misalignment of the spine. Nobody has knee problems without also having back problems. Perhaps by straightening the spine, limbs, and joints, these artificial replacements won't be necessary.

Often stroke and emphysema victims can't be helped. What would the effect on their illness be if these people could straighten their spinal column? I have no idea. Hypocrates, the father of medicine, said that we should look to the spine for the source of all illness. Perhaps then, there is reason for optimism.

Try the following experiment the next time you get a headache. Hold your arms in front of you as in the experiment with colors. Turn your head to one side and stare at the solid color red. Next make tight fists with your two hands. Sometimes, in about ten or fifteen seconds, your headache will disappear. It will reappear shortly after the experiment is stopped. Does this mean people with headaches should try this procedure? That is not for me to say. I am not a medical doctor, and not qualified to give such advice. But it is something reseachers should investigate.

CHAPTER 5

I believe that the most important human drive, assuming we have enough food to keep us alive, is conformity. We might vary slightly in dress, appetite, or work habits, but we are basically all the same. Twenty years ago we lamented that there were no longer any heroes. Today, we are told that it's better that way because teamwork is more important than individual effort. At least they waited until John Wayne died.

With that thought in mind the following contention might actually seem undesirable. I wonder if people could rebuild and straighten their spinal column, and if they could constantly straighten their limbs over an indefinite period of time, would they reach the age of one hundred and fifty years of age? Would the body work like a perpetual motion machine? If that happened, walking would be easy because each joint would exercise extreme elasticity. The heart would work fine because although large amounts of fat would have been consumed over the years, the body's machinery would be able to handle it properly so the arteries wouldn't get clogged. All the organs would perform as they should. Since the spine would be so perfect, and the body would not seem to work against itself, blood pressure would remain normal.

I don't know if these things could ever occur or, even if people could live to double the present life span, they would find it desirable. I talk to customers who are in their fifties about vinyl siding. I tell them the product has a fifty year guarantee, and their responses surprise me. They say "I'll be lucky if I live another twenty years," or "Who'd want to live that long?" People don't seem to mind death as long as they do not die of anything serious.

The next thought that occurred to me was that if people can live to be one hundred and fifty years of age, and we know that to be true because those people presently exist in some parts of the world, can people live to be several hundred years of age as mentioned in the Bible? I considered this possibility because more and more biblical pronouncments are being found to be true. I searched the book of Genesis for clues, and discovered what could be the answer. I was trying to investigate the idea that if our senses affected our spine and subsequently our health, then perhaps the sounds we uttered did the same thing. While reading the first book of the Bible, I noticed that the life span of the people mentioned was many hundreds of years. This occurred regularly until the story of the Tower of Babel. Immediately after that incident there is a geneology in which the life span of the people mentioned steadily decreased, until finally Terah, father of Abraham, is mentioned. He lived only one hundred and twenty years. This is not a long life by early biblical standards.

What caused the drop in life span after Babel? To find the answer I analyzed the Hebrew Language before and after the event. In doing so I make several assumptions. First, the words I read were the actual words of Genesis. It is supposed that we do not know the actual words of the Torah because it was originally written without vowels. It is one thing to say a language loses something in the translation, but to say that it also loses something in the original is farfetched. Besides, if we read the words as written, the theory works.

Secondly, the Ashkinaze pronunciation must be used. I believe the Sephardic pronunciation, the way Hebrew is now spoken in Isreal, and Reform, and Conservative Temples, was never the way the Hebrew Language was meant to be spoken. Thirdly, while pronouncing the words, the Yud without a vowel should be pronounced. Currently it is considered a silent letter.

What I discovered was that prior to Babel, all multi-syllable Hebrew words were written in such a manner that it was impossible to accent any syllable. To find a whole body of words so written is, in itself, unique. When a word is written in that way it is what I call a perfect guttural sound, as opposed to the throaty sounds of, for example, Modern English.

Monosyllable words were also spoken as guttural sounds. For example, the word, "Key", would ordinarily be a throaty noise. However, it becomes guttural when the supposidly silent, "Yud," is pronounced. Some words like, "Shame," are throaty, but when said with the preceeding word become guttural phrases. After Babel, as the language gradually changed and became less perfect, the human life span also gradually decreased.

One of the most facinating aspects of this phonetic detective work is that it explains why the Lord is sometimes referred to as, "Adonoi," and sometimes as, "Eloheim." The reason is simple. The words, when properly pronounced by duely emphasizing the, "Yud," sound are, "A doinoi," with the extra emphasis on the, "Oi," making it a word that cannot be accented, and the same for, "Eloheim," as it becomes, "Eloheeem." Now to see which word is to be used, it must be said with the preceeding word. When so done, these words cannot be interchanged, because by doing so it becomes phonetically impossible to pronounce the very necessary, "Yud," sound, which means the word goes from guttural to throaty.

An example of this is the first two words of the Bible. They are, "B-ray-chees Adonoi." If it was "Braychees Eloheem," the latter word could not be phonetically pronounced using the, "Yud," sound.

Sometimes the word for G-d is chosen because it makes it impossible to properly pronounce the subsequent phrase if the wrong word is chosen. However, this happens infrequently.

I now went back to my experiment of moving my hands back and forth. This time instead of doing so after performing work, I did so after speaking words. When I spoke a word that had a perfect guttural sound, my hands went forty five degrees each side of center. When the words were slightly less than perfect, my hands would go forty five degrees to the left and almost, but not quite, forty five degrees to the right.

When I said easy to speak English words, my hands didn't come close to going forty-five degrees to the right. When I spoke prepositions, or the word, "Aint", my arms barely went to the right of center. When I said the words, "Al, Mel, Joel, and Uh huh," my hands went all the way to the right, but barely to the left of center. This also occurred when I spoke the words, "what, when, where, and why," without pronouncing the, "H," sound.

22

There are people on this planet whose life span is well over a hundred years. They include the Hunzas, and the residents of the Steppes of Russia. We do not know why they live so long. Scientists visit them to try to find the answer to their longevity. They check their diet, soil, and environment, with inconclusive results. They just do not know the answer to this mystery.

Perhaps the researchers should analyze the language of these people in order to see how close it is to the perfect sounds of the very early Hebrew language. Maybe the physical effort exerted to constantly speak in guttural sounds based on a language whose words have syllables that cannot be phonetically accented, is the real physical effort needed to build health.

CHAPTER 6

Our civilization has official mysteries. Stonhenge is an example. Who built it? How were the stones lifted? What was its actual purpose. These are questions we often ponder, but may never know the answer.

We also have other mysteries which are so close to our daily lives that we fail to see in them anything unusual. Speech is an example. One would think that just as more and more aspects of modern life become complicated, language would develop into a more difficult and complex system of sounds. However, just the opposite has occurred. Words have become so easy to speak that we can communicate almost by grunting.

The Bible, the source of Jewish observance, makes no mention of covering one's head. Yet most Jews cover their heads during religious service. Nobody knows for certain why this practice developed. Some people even attach religious significance to the yarmulke, or head covering, and would never think of throwing one away. I believe the purpose of the yarmulke is that long, long, ago, the knowledge was imparted to the Jewish people, just like it was imparted to all other people, that head coverings were to be worn.

All great art and architecture, whether Eastern or Western, had a common denominator. It incorporated patterns of diversion. A person's eye always wandered as it viewed these objects, just like it did then viewing objects in nature. Modern architecture will not hold up as great throughout the ages because the viewer's eye is attracted to a solid color, and it is not allowed to wander. The automobiles of the nineteen hundred and eighties will never be classics for the same reason.

Early civilizations knew not to get only one hand hot or cold, and to do work with both sides of the body. While the Asiatics ate with chopsticks, the English drank tea using a cup and saucer. The saucer is now considered a sign of affectation and has practically fallen into disuse. Our English teachers insisted we never end a sentence with a preposition, and that we avoid the non word, "Aint." These are extremely throaty sounds. When people ate cheese, they ate it with crackers. When they drank whiskey, they drank soda or water. Mankind was also admonished to reverse its path. The warning was relegated to a superstition and ignored. The superstition said, "People should leave by the same door they enter." Inquiring minds should ask who gave this knowledge to mankind?

The point is that things were done for many years in the same manner. Even when people changed the meaning of the reason, forgot the reason, or could care less about the origin of the activity, it was done the same way. It is only recently we have changed the way we performed activity, and we are paying for that change with our health.

Other questions that interested people should want to know is if sounds and colors cause our arms to move when held a certain way, then what else will move?

How can this energy be harnessed? People who reported spotting unidentified flying objects didn't notice any recognizable type of propulsion, but did report seeing lights and hearing a humming sound.

On a more earthly plain, researchers should want to know why some people can feel their bodies move in the presence of certain sounds and colors, and others cannot. Interestingly, some people can feel the pulls involved with sounds but not color. I do not know why this occurs, but it is a phenomenon worth investigating. Perhaps inventors and scientists should build a machine which emits a wide range of sounds that affect the complete spinal column. The person with a deformed spine could then listen to the sound, feel the pull on his body, squeeze the appropriate hand and eventually feel his body straighten. Symbolically, this kind of treatment would be important because it would emphasize the fact that the patient's ability to help himself ultimately lies in his own two hands.

Our perception of health constantly changes. Several years ago doctors said they were conquering one type of illness, and now others were appearing. They said that these too would soon be cured. We are now told that this is as good as it gets, and we should be thankful for the miracles that modern medicine has wrought, because as everyone knows, "We are healthier than ever." I think that we are selling ourselves short.

I don't want to seem critical of the medical profession. Doctors see symptoms and they treat them. This is what they have done for centuries in every society. They operate, use drugs, or maybe even herbs, depending on the situation. They are doing a better job than ever. This is evidenced by the fact that people are living longer, more productive lives. The point that I hope I made in this book is that much of this pain and sickness that is now occurring, shouldn't exist. Whether hypertension, for example, should be treated with drugs, or more naturally with diet, exercise, or massive amounts of garlic doesn't interest me. Hypertension shouldn't occur in the first place.

CHAPTER 7

This chapter is included in the second printing of this book in 1996, which is six years after its initial publication. It represents my additional thoughts on the subject. My first thought is that fifteen years is too long to concern myself with anything. My ultimate goal is to wipe the subject from my mind completely. If I was merely feeling well, which thank goodness at this point of my life at age fifty-five, I am, it would be easy. But the pressure on my neck, which is fortunately diminishing, and my lying down either before or after dinner for long periods of time feeling various parts of my body laminating against the couch or bed with great force, are constant reminders. That is why, in a burst of vanity, I formed the Ethan Sherman Foundation. Its purpose is to raise money to give grants to scientists and researchers to see if my procedures and theories have a basis in scientific fact. The foundation is a 501(c)(3) public foundation.

I heard a famous television personality comment about the fact that the body breaks down after fifty. I'm knocking on wood with one hand and throwing salt over my shoulder with the other, as I type this, because at present, this has not happened. I take no medications, see no medical doctors, and except for the slightest touch of arthritis in my thumb, have no pains. My blood pressure is usually 115 over 70. I like to think that my program has something to do with it.

During the past six years the apparent twist in the middle of my neck worked itself out. That pressure, which is much less severe, seems to exist about two inches higher. I can't tell exactly what is occurring because I won't have more X-rays taken, but the area feels like it is twisted, compressed, and slightly leaning to the right. The activity that takes place in my neck seems to be reversing that situation.

I was a little prophetic in the first printing. Cancer, especially breast cancer, is reaching epic proportions. Cardiologists are seeing patients who are younger than in the past. The government is warning the public about the possible dangers from electromagnetic waves. Female flight attendants are experiencing increased instances of breast cancer and scientists admit they are puzzled as to the cause. I believe that I know why. It is exposure to the sounds that some appliances and motors make, including jet and diesel engines. These sounds cause the tuning fork quality in the spine to break down and ultimately this leads to a variety of illnesses.

During the past several years there has been a change in two items that I believe no longer makes them harmful. They are refrigerators and high speed burners found in modern furnaces. However, the new sounds of many computers, including lotto machines, and the sound of stereo television, cause the body to move, and that is why I think that they could cause health problems.

I received a small amount of feedback from the first edition. Many people felt that it was interesting, but didn't understand my point. Rephrasing many things would be impossible to do in a few paragraphs. Yet it is important for me to get the point across that it is not necessary to have the best possible spine. In order to do so we wouldn't play sports as we know them and life would be pretty drab. We would brace ourselves with one hand when we turned on a light switch, and we would compensate on one side as we did such routine activities evenly as reaching for food, or pulling up our pants, to name only two. The spine is meant to be far from perfect and let the person live a normal pain-free existence. On the other hand, by knowing the principles of how it works and under medical supervision, perhaps people could affect enough improvement by reversing wrongful activity to get relief from pain and illness.

Some miscellaneous thoughts about the subject are as follows: originally I countered a force caused by color, sound, or smell by squeezing the opposite hand. These forces are so strong that I found that in addition, standing on something with the opposite foot works for me even more effectively. Many common activities cause the spinal column to be less than perfect. These include jogging, and drinking out of a can or bottle with a straw or a glass filled with ice. These are minor offenses. One activity that is more significant, in my opinion, is consuming hard candies. By doing so and moving the arms back and forth I believe people will notice their arms go to the right. Countering this activity by standing on something with the left foot and squeezing the left hand greatly affects the hip area.

As my spine untwists and stretches it seems to move slightly to the right and then straightens. I noticed this when I was shaving. While my body often moves when I sleep or rest, it does so with greater force when lying on a hard surface such as a floor or an elevated plywood board.

I initially wrote that the sense of smell has a minor impact on the spine. It is, in fact, significant. Most things that emit a smell will affect the back when sniffed closely. Garlic, onions, and fresh ground coffee are a few examples. The former sends the arms to the right and the latter two sends them to the left. Many flowering weeds or plants do the same. Surprisingly, tree leaves, which give off no odor, do the same thing. Maple leaves send the arms to the right and oak, walnut, or catalpa send them to the left. Counteracting these forces caused a significant stretching of my spine in the area of the lower back. Just like staring at colors, when I stopped sniffing things, the stretching continued.

My last discovery is one I write about with some reluctance. I may have crossed the line between the unusual and the weird. After some consideration I figured, "Why stop now?" It is possible to move the spinal column with the mind. This is done in the following manner. Stand, as I previously described, and loosely swing the hands and arms back and forth, noting they act as a meter going about the same distance each side of center. Next, ask yourself questions requiring a "yes" answer. Notice the arms go to the right. Next, ask yourself questions requiring a "no" answer and notice they go to the left. By countering the forces by squeezing something in the opposite hand or squeezing the opposite hand and standing on something with the opposite foot, the chest area was affected. I would find myself, when lying down, getting comfortable with my hands and wrists together under my chest and that part of my body pushing down on them as if they were a fulcrum. This experiment or exercise has great significance beyond working with the spine. I am including an article about it that was submitted to a couple of magazines but never published.

CONTACT YOUR GUARDIAN ANGEL: IT'S REALLY QUITE EASY

I believe that people can find out if they have a guardian angel, and if they do, make contact with him or her. While working on a project that had nothing to do with angels, I made this unique discovery.

The body has a built-in meter. It points to things and moves in ways never before mentioned; at least I have never heard in the past of anybody doing what I did. Here's how it works. Stand up, take a deep breath and relax. Place your elbows by your side and extend the rest of your arms forward. Place your hands together. Do not clasp them, but have your fingers touching fingers and palms touching palms.

Next, move your touching hands and arms from side to side loosely like a pendulum, noting that they go approximately forty-five degrees each side of center. Now ask yourself out loud, or even to yourself, questions that have a "yes" or "no" answer, all the while moving your arms freely back and forth.

Notice that if the answer to your question is "yes", the arms will swing about eighty degrees to the right and only ten degrees to the left. Conversely, if the answer is "no", they will swing approximately eighty degrees to the left and only ten degrees to the right.

What I did after discovering the above is ask myself if I had a guardian angel. My arms went mostly to the right, signifying the answer was affirmative. Next, I asked if it was male and my arms went to the left. When I then asked if the angel looked after others, my arms swung to the left. Then, I asked if it was a relative and the answer again was negative. I then asked if the angel was at one time alive and my arms went to the right. Then I asked if it was a person who lived in the twentieth century. The answer was again negative as it was when I asked about the nineteenth, eighteenth, and seventeenth centuries. When the sixteenth century was mentioned, the arms moved significantly to the right.

When I asked if it was okay to write this article, my arms also went almost all the way to the right. I have repeated the above experiment several times and always get the same results.

Next, I pondered about what source I had tapped. I am aware of the Biblical warnings about consulting fortune tellers and mystics. When I heard of warnings about ouija boards I quickly got rid of mine. However, I do not think this is a harmful source. This is based on several reasons. First of all, I asked it and my arms went all the way to the left. Secondly, I asked it about questions relating to morality. "Should I pray?" evoked a positive, while "Should I steal?" evoked a negative response.

The following experiment was quite interesting. Jewish dietary laws

28

forbid eating shellfish, pork products, the hind quarter of meat, or mixing meat and dairy products. When I moved my arms back and forth while visualizing the above or actually eating of the above, my arms moved to the right when spare ribs and shellfish were involved. They moved to the left when the experiment included the mixing of meat and dairy or eating the hind quarter of meat. Other people duplicated this experiment.

Finally, I don't believe this is a harmful source because I feel that whenever one source of energy is available, a counterweight also exists. If sorcery is considered evil according to the Bible, then a force for good must exist on the same plane. Perhaps this is it.

I believe others can repeat my experiments if they do so desire. What it all means, I cannot say with certainty. Perhaps nothing. I don't dwell on this nor do I trivialize it. In other words, I don't seek the results of horse races.

It is very difficult to write about anything considered "alternative healing" and have it taken seriously. At best it is usually damned with faint praise. There is an "It's good because it works for you" attitude. Yet, without understanding how the spine works, there will never be true health. We will witness so much tinkering around the edges in the form of outlandish diets, massive exercising, etc., that most people will not be able to follow expert advice even if they wanted to do so. Furthermore, the results obtained from following this advice will at best be marginal.

We have to realize that the spine is as great a part of the body as any other, and it is being abused in ways we cannot imagine. When that happens, there will be such a significant drop in illness that writing about it in newspapers and magazines won't be worth the space.

CHAPTER 8

This chapter was written in 2003, seven years after the previous one, and it represents my final thoughts on the subject. My first thought is that twenty-one years is a long time to be concerned about anything. On the other hand, when I was younger, I knew that I wanted to become involved with something that greatly held my interest; this fulfilled that requirement.

During the past seven years there appeared to be a rise in illness. I cannot quote statistics, but it seems that Heart Disease, Cancer, and Diabetes have become epidemic. There are also increases in Autism, Alzheimer's Disease, and Asthma. We are seeing knee pain in young adults, and, very recently, inexplicable sudden death. Parkinson's Disease and Multiple Sclerosis are on the rise. The poster people for these diseases are celebrities in their forties; yet experts tell us that it is because people are living longer. The two hospitals in my city have almost doubled in size; yet the area population has only moderately increased.

Almost everything is blamed on lifestyle. I don't think that this is so. The Western diet has served us well for centuries. I do not believe that all of a sudden it caused our health to implode. In fact, after blaming diet for almost every ill the experts are still pretty much undecided as to what we should eat. At one time people were told to eat whatever they wanted. The theory was that individuals would chose the correct foods because they could sense what their bodies were lacking. Now we need experts to tell us what to eat.

I believe that people should avoid a sedentary life style. If they enjoy exercise they should do it. However, my beliefs about the health effects of exercise are the following: We are getting prematurely sick because of the breakdown of the spinal column primarily due to the sounds to which we are exposed. It is a problem that can be solved at the manufacturing level. Once that happens illness will drop to pre-World War II levels. Magazines and television will seldom report health issues because they will not be of much interest to the general public. Health costs will again become out-of-pocket expenses. This could very easily happen. Nevertheless, since we are not going to do that in the near future, we have to fool with the symptoms; so if the medical authorities say that we have to vigorously jump around for thirty to sixty minutes a day to lessen the chances of becoming ill, then who am I to contradict them? It is a crude approach to good health that doesn't even work that well, but it is the best we have. Furthermore, being forced to exercise that much robs us of our freedom, and is also very hypocritical. To be told that it is our fault that we are becoming prematurely sick when something harmful is actually being done to us is wrong.

I have a theory about Obesity and Diabetes; this is just a guess, but I don't think we are getting Diabetes because we are becoming fat. I think we are becoming fat because we are getting Diabetes.

Since I began my foundation in 1995, I have been unsuccessful in carrying out its objectives. Perhaps Ethan Sherman collecting money for The Ethan Sherman Foundation seems suspect. However, what bothers me is that I can't get scientists to do my proposed research, even when I offered my own money to get the projects started. Telephone calls and letters to scientists and research institutions go unanswered. One day I went to the Science Department of a local college and told the girl behind the desk that I wanted to see a scientist. She said that there was one in an office around the corner. I meekly and respectfully knocked on his door, introduced myself, and told him of my proposed research. He looked at me with disdain and said, "So do it." He didn't say, "So what do you want from me?" He might have added those words because it was implied in his tone.

At least I got my answer. Researchers just don't talk to non-researchers about science. I was really stunned. I talk to everybody, and it surprised me to be treated like that. About a week later my suspicion was confirmed when I heard a radio news item which said that scientists at Yale who are on the fast track cannot even talk to non-scientists about that subject.

Actually the research I read about during the past several years supports my conclusion. A case can be made for unique sounds harming health by merely reviewing existing studies: Researchers are looking for an environmental cause of Cancer. The sounds that I mention which exert this force on the body are an environmental cause. The Federal Government has a whole department that is looking into the possible harmful effects of electromagnetic waves. It gives huge research grants to study the possible harmful effects of low level radiation. The Federal Government almost has it right. The products are indeed harmful; but it is the unique sound emitted, not the radiation or electromagnetic waves that is the problem.

There is a lot of illness associated with flying. Research has shown that female flight attendants have high rates of Breast Cancer. Pilots have high rates of Leukemia. There is Phlebitis associated with long flights. There are neurological difficulties blamed on altitude. When people aren't getting seriously ill, they experience air sickness. When airports are built or runways are extended, there is more Cancer on the ground. I seem to notice that about a week or so after people take long flights they coincidently come down with an illness that appears completely unrelated to flying. I believe all these health problems are related to the sounds of the jet and turbo-prop engines. This is the most compelling evidence, and should be the Rosetta Stone in the war against Cancer.

Other research showed that people who commute by rail three hours per day or more experience a change in heart beat. Stress was considered the problem. This cannot be true because there is nothing that is more fun than riding on a train. More interestingly, it was noted in another study that people who slept during their train trips experienced a rise in blood pressure. No explanation was given. I believe the reason for these changes was due to the sound of the ventilation system located under the commuter railroad car, whose sound permeates the rail car. This sound exerts a noticeable force on the body. Amtrak's older coaches have a different ventilation system which does not exert this force; but their sleeping cars and morel modern trains have a newer system that I consider harmful. Interestingly, you do not hear of people who travel by Amtrak coaches getting Phlebitis; yet they do on shorter airplane trips.

A few years ago a study was done on the so called Black Swagger. It stated that young African American males walk differently than they did in the past. The reason given by sociologists was that these people were making a statement. I believe the cause of the Black Swagger is that young African American males continuously listen to FM radio played at a very loud volume. The FM signal exerts a force on the body which alters the spinal column and subsequently their walk.

After 1990 research showed that there were unexplainable drops in Cancer and Heart Failure rates. Sudden Infant Death rates also dropped; but that was credited to the fact that babies were being put to sleep on their backs. This is what I believe really occurred: One of the most harmful sounds comes from the frost-free refrigerators which were built from about the early 1970's to about 1990. After 1990 the refrigerator motor was changed so that its sound no longer exerted this harmful force. However, from 1973 to 1990 (when its sound exerted a force) the NCI said Cancer rates increased 1.2% per year. I cannot overstate the harm that I believe was—and still is—caused by these terrible machines.

There are a lot of health problems in the African American community that are often attributed to racism. I do not think that racism is the cause. Black families are moving into city homes. Usually these houses have eat-in kitchens equipped with the older refrigerators. This is causing them terrible physical problems. Sadly, these pre-1990 refrigerators never die.

Power line workers have high rates of Leukemia. The on-the-job sounds that exert a force on their bodies come from the hum of defective street lights and the highest capacity power lines. In addition, they are exposed to the sounds of the diesel engines and maybe even (I haven't checked it) their radios.

Farmers who do their own farming as well as farm workers have high rates of Cancer. The farm chemicals are blamed for that problem. Maybe they are the cause; certainly they are not doing the person exposed to them any good. Nevertheless, I would not rule out their constant proximity to the sound of the diesel engines that power the farm equipment.

I want to update my list of sounds which exert a force on most people' bodies even though I am being somewhat repetitious. They come from the:

Diesel Engine	All Radio Signals I checked other than AM
Jet Engine	Heat Pumps
Turbo-prop Engine	Central Air Conditioning
Supermarket Refrigeration	Refrigerators built between the early 1970's
Soda Machines	to about 1990
Electric Water Heater	Many Copy Machines

32

Humidifiers	Most Forty Pint Dehumidifiers
Air Purifiers (many)	Modern Cash Registers (many)
Sirens	Air Horns
Impact Tools	Water Fountains
Bathroom Hand Driers	Hair Driers (some)
Bathroom Ventilation Fans (some)	Modern Elevators
Drives of the fastest Computers	
Modern, more powerful Vacuum Cleaners	

The high speed residential furnace no longer makes a sound that exerts a force on the body. However, several extensively used modern products do:

- The stereo television sound exerts a force, while the non-stereo television does not.
- The sounds that come from DVD's, play stations, music piped into stores and malls from satellites do.
- I think (but I am not sure) that satellite radios and the newest sound in movie theaters do.

To test this I must stand as I described, with my elbows near my side, arms extended, and fingers hanging limply. It's bad enough that this looks silly. However, there are now signs placed all over that say, "Report anyone who looks suspicious." Therefore, sometimes I can't check out things as thoroughly as I would like.

Modern postal scales and the newest supermarket checkout equipment now have a sound that exert this force. I don't want to change the chatty and cheerful tenor of the conclusion of this book (for me this is chatty and cheerful) and sound too ominous, but these unique sounds that exert this force on the body should be cause for concern.

As far as science is concerned, we are faced with a double edged sword: People are living longer than ever, and we have science to thank. I have heard people in their eighties exclaim that they never expected to live as long as they have and they are indebted to their doctors for this. When doctors see a harmful symptom they prescribe a pill to correct it. Early detection and marvelous operations are wonderful tools for curing sickness and increasing longevity.

Nevertheless, science has almost completely misjudged the cause of the massive amount of illnesses that we suffer. I hope my book and foundation can correct this situation, although I realize that I might be tilting at windmills.

The research that I want to sponsor through my foundation is the following: One group of about twenty-five mice would be exposed to twenty-four hours of television listened to on a stereo television. Another group of twenty-five mice would listen to (they could actually watch the shows if they wish) the same programming on a non-stereo television. No animal would be euthanized unless it became ill. The droppings of both groups would be regularly analyzed; life spans and behavior would be noted; a pathology would be performed on all deceased animals. (Incidently, it was recently reported that people who watched television extensively had higher rates of Diabetes. Naturally, they had to somehow tie this into diet rather than correctly placing the blame on the unique sound of the stereo television.)

I believe that the results of this experiment would shed light on the cause of a number of illnesses and establish the unique sound of stereo television and other products whose sounds exert a force on the human body as the major factor in the dramatic rise of many illnesses.

Another research project that I want to sponsor is to have physicists take a hard look at the way the body moves when we stare at colors at angles and in the presence of various sounds. Is this free energy? Can it be harnessed in other ways beside its connection to the spinal column?

Scientists are bound in their projects to only build on existing research. A nineteenth century scientist named, Lenard, said that color gives off energy. My experiments showing how color actually moves the body builds on that theory, so I was hoping that physicists would want to investigate this phenomenon more thoroughly.

A third research project I have in mind is to phonetically analyze the Book of Genesis in Hebrew to see if it was written differently before and after Babel. As I previously mentioned, biblical people lived very long lives until Babel; after which the life spans dropped immediately and dramatically. If before Babel no word could be phonetically accented; and after Babel accents crept into the language, and the life span dropped, then speech-not-genetics-could be responsible for extreme longevity.

I feel that this is a tantalizing theory. I was told by a rabbi that I am wrong. There is something called tropes which indicates which syllable is to be accented. I believe that while it is physically possible to say one or many syllables louder than another, the actual force of the other syllables is such that they carry equal weight.

Finally, I believe that staring at a color at an angle, then making fists and pushing against that force does a tremendous amount of physical good. I want to give grants to doctors to perform clinical trials in addition to the conventional treatments: When my mother was ninety-two, her home health nurse described her heartbeat as "squishy." I had her stare at the color red placed off to the side for thirty seconds while clenching both fists by her navel with napkins rolled up in each hand for extra clenching power. After that time elapsed the nurse listened again to her heartbeat and said that it was beating normally. This improvement was only temporary.

My father had a stroke when he was in his late eighties that affected his left side. He could only take small steps and had to have assistance when he walked. He did my procedure twice a day for thirty seconds each time. At the end of the week he stopped because his neck started to hurt, and he didn't want any part of it. However, I watched him carefully and noticed that when he sat on the couch and watched television his ankles would be wrapped around each other with enough force so that they would become red from the pressure. In addition, there were marks on the back of the lower part of his leg as it pushed against the sofa. This occurred unbeknownst to him. I believe that the constant pushing of body part against body part and body part against furniture would have eventually enabled him to walk unassisted.

Another family member faced dialysis. I had her try this procedure twice a day for thirty seconds each time for two weeks. At the end of that time she still had to go on dialysis. In addition, she was very upset because her neck and even back started to hurt, and she never had those pains before. After she stopped, her neck and back continued to bother her for almost two months. Yet while her neck was hurting and movement was occurring, she handled the treatment very well. There was a marked difference between the way she left the building and the way the others did. Since her neck pains ceased, she now struggles with the dialysis treatment.

One of my customers is diabetic; I stopped by her house as she was taking her blood sugar reading. I had her stare at colors and make fists twice within a ten minute period and then retest. The reading dropped by almost ten percent.

It would be interesting to see the results doctors would see if a number of their patients with a wide range of illnesses tried this procedure in addition to their conventional treatment. This would mean that they would have to collaborate with chiropractors. In my case the chiropractor basically does two side-to-side adjustments and moves my neck back and forth. This is done every month.

In the meantime I'm content that at sixty-one years of age I have been able to so far avoid doctors; (except for my four cases of Lyme Disease) I hope it continues. When the pressure on my neck gets too uncomfortable, I lie down. This is the way I watch television in the evening. I only have to snap my neck three or four times a day, and things are manageable. I still believe that I have a very severe rotation in my neck that is just taking a very long time to work itself out. As my neck untwists my chest and rib cage expand so that I have a larger upper body than I did previously.

Sometimes when I am lying down I can only get comfortable with my arms placed under my rib cage. They then exert an upward pressure so that the area between my hips and rib cage stretches. Over the years various parts of my head and face have pushed - and continue to push - against the arm of the couch with great force, leaving red marks on my forehead, temple, and cheekbones, giving them a chiseled appearance. Often people tell me that I look well, or that my appearance has greatly changed. In the beginning I would tell them what I was doing. After a while I got smart and changed my response to "Thank you."

There is a prevailing view in this country that anything that harms health is one more thing about which we have to worry. Conversely, anything that we do to improve our health is one of many things that contributes to a healthy lifestyle. I think that this attitude is due to deliberate brainwashing by the media. Everything has to be one of many: A great actor cannot be another Dustin Hoffman. He must be another Dustin Hoffman or Jon Voight; a great place to vacation can't be Bermuda. It must be Bermuda or the Bahamas.

I believe that my theory about the decline of health being related to the destruction of the spinal column is much more important than that. It is not merely one of many.

Finally, in the previous chapter I described a method of contacting your guardian angel. Don't fool with it! Nothing terrible happened to me, but a clock and decorative plate mysteriously fell off of the wall when no one was in the room. Actually, the clock fell the day after I rehung it following the change of Day Light Savings time. Perhaps I didn't hang it properly; but you never know....

www.ingramcontent.com/pod-product-compliance
Lightning Source LLC
Chambersburg PA
CBHW070238290526
45789CB00004B/1682